HOW TO UNDERSTAND E=mc²

Christophe Galfard has studied Advanced Mathematics and Theoretical Physics at Cambridge University, England, where he did his Ph.D. on black holes and the origin of our universe under the supervision of renowned Professor Stephen Hawking. Praised for his ability to explain difficult ideas with simple words, Christophe has been, for the past few years, devoting his time to spreading scientific knowledge to the general public. He has given public talks in front of hundreds of thousands of people, children and adults alike, throughout the world. He is a regular guest on TV and radio shows in France, where he is one of the most acclaimed popular science writers and speakers. Christophe has written many award-winning popular science books for children about the Solar System and our Earth's climate before writing *The Universe in Your Hand*, his first book for adults, now an international bestseller translated into 20 languages.

HOW TO
UNDERSTAND E=mc²

CHRISTOPHE GALFARD

Quercus

First published in Great Britain in 2017 by

Quercus Editions Ltd
Carmelite House
50 Victoria Embankment
London EC4Y 0DZ

An Hachette UK company

A CIP catalogue record for this book is available
from the British Library

ISBN 978 1 78648 495 6

CONTENTS

Foreword

$E = mc^2$.

E is for Energy.

The same kind of energy that makes your car run, your light bulbs shine, your fridge hum.

m is for mass.

The same kind of mass you and I, the air, the seas, mountains and clouds and all the known matter in our universe, are made of.

And c^2 is the square of the speed of light.

A huge number by all means.

$E = mc^2$ says that energy can become mass. And mass can be turned into energy. An awesome lot of it. It tells us why we can split the atom and how stars shine, and even how nature can create particles out of nothing. But that is not all.

$E = mc^2$ is a beacon of sorts, a signpost indicating the entrance into a new reality where not only mass and energy, but also space and time have meanings that are not the expected ones. It has implications in the realm of the very small, and the very big. So much so that it shaped pretty much all of the twentieth century, including how we think about ourselves, leading to the world we live in today.

PART I:

Light

CHAPTER I

Historical Introduction

Around the beginning of the twentieth century, almost everything that was scientifically known about reality was based on what Newton had summed up from past knowledge, and discovered himself, some 180 years earlier. It corresponded to how our intuition tells us nature behaves.

But this was about to change radically.

After all, for the past ten thousand years, our bodies have hardly evolved. We've had pretty much the same eyes, ears, fingers, tongues and noses in all that time. That makes us all equal at birth, throughout the ages, when faced with trying to understand what is happening around us.

Thanks to centuries of questioning, wondering and techno-logical improvements, the beginning of the last century saw our species reach a new level of awareness. We realized that the laws of nature we intuitively believed to be true everywhere, throughout space and time, were not the ones we thought.

Compared to the immensity of our universe, we are tiny.

Compared to the minuteness of fundamental particles and their quantum world, we are huge.

We float in between these two infinities, one large and one small, and our senses only allow us to probe what is around us in a limited way.

About 100 years ago, we saw that as one drifts away from the safety of our scale of reference, the laws of nature begin to change – drastically. What we experience on a daily basis is but an approximation of realities our senses are not made to detect. This knowledge is what makes us different from all the humans who have lived before us.

As of today, we know of three paths that lead to unforeseen aspects of reality. One is the big. One is the small. And the last one is the fast – the realm of high velocities.

Just as it is true that we are neither large (compared to the universe) nor small (compared to particles), it is also the case that we never move fast. Even the fastest rocket ever launched is pretty much a slug in comparison with whatever flies at the speed of light.

But wait, doesn't light travel instantaneously?

I know you know it doesn't. It travels at a particular velocity which we call the speed of light. Scientists refer to it as the letter 'c', for celerity, which means swiftness. And if it has been granted the honour of a letter, which neither your velocity nor mine will ever achieve, it is because there is something really peculiar about it: in a vacuum, light always travels at the same speed.

Always. Independently of who is measuring it.

This is half the reason why $E = mc^2$.

And to see how this was understood, we need to begin by measuring the speed of light.

CHAPTER 2

The Speed of Light

Picture yourself in a dark room.

Your hand is on the light switch.

You are totally focused, because you are about to figure out the time it takes for light to travel from the bulb to your eye.

You turn the lights on.

But you don't detect any delay.

As far as your senses are concerned, the room was lit instantaneously.

Galileo tried such a trick as far back as 500 years ago with a source of light about a mile away from where he stood, and wasn't able to notice any delay either. Yet there is one. Around Galileo's time, no device was precise enough to detect it. To be able to see anything at all, our ancestors would have needed light to travel through much, much greater distances than anything that can be found on Earth.

And in 1676, that is exactly what Danish astronomer Ole

Rømer did. He studied Io, one of Jupiter's largest moons.[1] Like most planets, Jupiter does not shine on its own. It is lit by the Sun. So there is a shadow behind it. Io moves in and out of this shadow on a very regular basis, making it pop in and out of the darkness. Thanks to Galileo's newly invented telescope, Rømer noticed that it took more time for Io to disappear and reappear when the Earth was moving away from Jupiter than when it was moving towards it. For Rømer, that was it. A telltale sign that light did not travel instantaneously. He even estimated how fast it was to within about 20 per cent of today's value. Not bad at all for a first attempt.

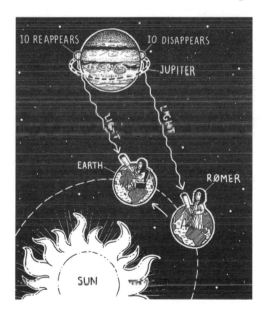

About 200 years later, around 1860, the physicist James Clerk Maxwell, from Scotland, started a series of scientific revolutions which led to no less than the science of the twentieth century. In a time when people rode horses to travel around and used candles to work at night, Maxwell discovered that electricity and magnetism were two aspects of the same phenomenon – electromagnetism – which, when perturbed, gave birth to a wave.

Just as a moving buoy on a lake creates waves on the surface that move away from the buoy at a certain speed, if you move a magnet around, you get a wave – an electromagnetic wave. That is what Maxwell's equations implied. And he, of course, wondered how fast these waves travelled. Experiments told him: at the speed Rømer had found. The speed of light. Maxwell did not believe this was a coincidence. As strange as it may sound, he had discovered that light was an electromagnetic wave.

But this led to a new puzzle.

A wave on the surface of the ocean travels on water; a sound wave travels through matter.[2] But what does a wave of light travel through? In a room, we do see a candle burning on a table and we do see faraway stars in the night sky. In a room, there is air. In space, there is nothing. Nothing that can be seen, anyway. And yet, light travels through both. Following Maxwell's discovery, scientists imagined there had to actually be something out there in outer space, and here on Earth, that we could not see. Some sort of medium that would fill the entire universe, a medium that could wiggle and welcome the passage of a wave. A wave of light, that is. That medium was called the *luminiferous aether*, or *aether* for short. Hardly any of the most brilliant scientists of the time doubted its existence, but if you've never heard of it, don't worry, that's normal. It doesn't exist.

Picture yourself on a ship, at sea. It is windy, you are sailing fast. Another ship is behind you, sailing at the exact same speed. She blows a horn to say hi. Carried by the wind, the rumbling sound of the other boat's horn reaches you faster than it would have on a calm day. Politely, you blow your horn back, but your acoustic signal now has to fight against the wind.

Comparing the two travel times gives you a way to estimate the speed of the wind.

In the 1880s, two American scientists named Albert Michelson and Edward Morley did exactly the same experiment to detect not a wind of air, but a wind of aether. The ship they used wasn't a random boat on the ocean. It was the one we all board at birth to travel across the universe: the Earth itself.

The Earth completes an orbit around the Sun in a year. And it so happens that the orbit is rather round, meaning that we move pretty much in a circle around our star. So whatever day it is now, as you are reading this, you, me, everything here on our planet is moving towards faraway stars that lie in the exact opposite direction to those we were moving towards six months ago, or will be in six months' time. That is what happens when you travel in a circle.

Now, even though we do not notice it, the Earth travels quite fast around the Sun: at about 100,000 km/h. So if there was some aether around, filling everything and blowing in some direction, there should be a 200,000 km/h speed difference between how fast the Earth moves, at six-month intervals, with respect to the aether.

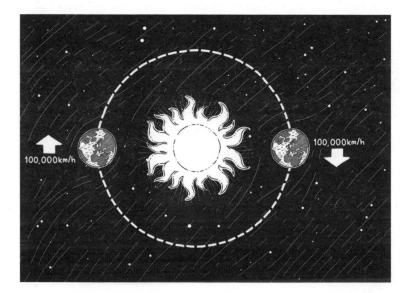

Michelson and Morley's experiment consisted basically of firing light rays towards the same faraway stars out there in space every six months. The Earth would first be moving towards one of these stars, and six months later away from it. If light travelling through the aether was at all like sound travelling through air, light's travel time between two spots should change and give away the speed of the aether wind.

But they found no difference at all. Zero.

No wind. That was unexpected but (most) scientists got over it.

However, they could not detect the speed of the Earth either. This 200,000 km/h difference did not appear at all.

That came as a real shock, and not just because it meant there was no such thing as an aether.

To understand why, picture yourself standing on a cricket pitch. You throw a cricket ball, aiming at the wicket.

Now throw it again, with the same force, while on a rocket travelling towards the wicket at 200,000 km/h.

You wouldn't expect the two cricket balls to hit the wicket at the same velocity, right?

Well, rather unexpectedly, that is exactly what Michelson and Morley discovered about light.

Whatever its source is doing, light always travels at the same speed.

Exactly 299,792,458 metres per second.

'Exactly' because, since 1983, that is how a metre is defined: it is the distance travelled by light in a second, divided by 299,792,458. There's nothing to argue about there.

Light, as we have understood it ever since, is an electromagnetic wave that travels not through aether but through, well, nothing. And its speed, in the vacuum of outer space, is constant[3] – c – whose value is 299,792,458 m/s.

Once this was known, humanity was just one principle away from $E = mc^2$. A principle that Galileo had first stated, centuries ago, and which Einstein modernized. It is a way to think about the world from the point of view of a moving object.

PART II:

A Theory of Moving Objects

CHAPTER I

Everything Is Relative

Michelson and Morley had found that whether you move or not when switching on a torch, it makes no difference to the speed of the light you shine. In other words, as far as light is concerned, speeds don't add up.

Albert Einstein may not have been aware of this experiment. But even if he was, he probably didn't care much about it. He fancied trying to picture everything in his mind. Experiments came a long way behind. So if he thought about this speed of light conundrum for a long time, it is not because of Michelson and Morley's strange result, but because it had already appeared in Maxwell's electromagnetic equations: the speed of light was in the electromagnetic wave equation Maxwell had found, and it was a constant.

Faced with this, Einstein came up with two ideas, or principles, that he believed nature should somehow follow to account for what he believed reality should be.

His pondering was led by the following question: could there be a special movement that would make one see reality in a more fundamental way than any other movement? Does the universe and its laws look simpler if you are lying in bed, or travelling in a car, or aboard a spaceship in outer space?

Galileo Galilei first thought that it was not the case 400 years ago. He realized that were you to be inside a windowless cabin on a ship at sea, no experiment would ever be able to tell you whether the ship was moving or not. You can try this today on a train: haven't we all been at some stage fooled by seeing a train leave the station through the window, thinking that our train was leaving instead? Even in a car driving at a constant speed: close your eyes (if you're not the driver, obviously) and try to feel the car's velocity. You'll see that you can't.

Isaac Newton and French mathematician and physicist Henri Poincaré agreed too, and so did Einstein: if you are in a closed box, with no window to tell you what's going on outside, no experiment will ever tell you whether you are moving or not. And, according to them, this should actually hold everywhere, not just on Earth.

Visualize yourself in a spacesuit in outer space, far away from everything. Your helmet allows you to see what is around you, but not the faraway stars. You are completely still.

You suddenly spot two human-looking shapes flying at

speed towards you. They are wearing strange spacesuits and their helmets are hiding their faces. They may be aliens, for all you know. Thrilled, you wave at them, but they whiz past you tremendously fast. Just like you, they don't have any rocket or anything to propel themselves, but boy do they move quickly! Impressed, and slightly jealous, you contact them via your radio set while you still can, to pay your respects. To your great surprise, they reply that they too were delighted to have spotted you, but that as far as they're concerned, they're not moving at all. *You* are. Tremendously fast, by the way.

And they are right.

And so are you.

This became Einstein's first principle: as long as one does not accelerate, it is impossible to tell who is moving and who is not because there is no absolute frame of reference to refer to. There is no luminiferous aether, no current against which one could measure an absolute, objective speed. Only relative speeds are meaningful.

Because of that, the speed at which you are moving does not make you special, and hence the laws of nature should be the same whatever your velocity. Following Galileo, Newton and Poincaré, Einstein called this the principle of relativity. In mathematical language, it can be stated like this: if a law of nature writes A = B for someone travelling at some given velocity, then it should also be exactly A = B for everybody

else who is also moving at a constant velocity, however slow or fast. Not A = something else.

The second principle is that in a vacuum, light always travels at the same speed.

To see that these two principles lead to a new vision of reality, let's do what Einstein did: a thought experiment.

Picture yourself sitting on a chair made of light, shooting through space at the speed of light.

You are on your way to a very futuristic date taking place on the other side of the galaxy.

Slightly anxious about your looks, you take a mirror out of your pocket and hold it in front of you to check your hairstyle. Now, the question is: do you see yourself in that mirror? A very important question indeed.

For you to see yourself, you'd need to have some light travel faster than you, otherwise it never reaches the mirror and can't bounce back towards you. You'd need to fire some light travelling faster than light. And that is not an option. So the answer is no, you would not see yourself. Actually, the real reason why you wouldn't, as we understand it today, is more unsettling than that: at the speed of light, your time would stop.

Your heart would not beat.

Your cells would not age.

Your clock would not tick.

Time would be frozen and distances shrunk. You wouldn't see yourself in the mirror because you wouldn't see anything at all (nor be able to pick up the mirror in the first place). Still, you'd be travelling fast and shooting past marvellous places that you would not be aware of.

Disappointing, I know.

But what might be even more disappointing is the following: in truth, you'd never be able to make that trip. For you to travel at the speed of light, you'd need to be made of light too, not matter. Massive particles (and you are made out of plenty – no offence intended) cannot reach the speed of light. Ever. Only massless particles can. That is a consequence of $E = mc^2$.

These are but two of the many strange consequences of Einstein's two principles above. They mean that nature's behaviour at high velocities forces us to reconsider many things we once took for granted.

This is part of what Einstein laid out in 1905 in a paper called 'On the Electrodynamics of Moving Bodies'. He was a completely unknown 26-year-old patent officer in Switzerland at the time.

A few months later, following on from that idea, he saw that one also had to reconsider the meaning of mass.

This is what led to $E = mc^2$.

A genius like no other was about to start reshaping the world.

CHAPTER 2

To Agree to Disagree

You are back in space, in your spacesuit, remembering the two astronauts who flew by you a minute ago. You still haven't seen their faces and start wondering: aliens or not, could it be that the laws of physics they experience are the same as yours? And I mean *exactly* the same, i.e. if you find a formula to describe a phenomenon, would exactly the same formula work for them as well, to describe the same phenomenon, as it is seen from their point of view, even if they are very fast moving and aliens?

The principle of relativity says that it should be the case.

But we already have a problem.

Maxwell's understanding of electromagnetism was, and still is, extraordinarily successful. To give you an idea, it brought us radio sets, television sets, radar, mobile phones, microwaves and most other such electronic tools. It also allowed us to look into the night sky using electromagnetic waves that our eyes cannot detect: X-rays, UV radiation and so forth. Quite something. It baffles me that Maxwell is not better known by the general public.

But here is the problem: Maxwell's equations don't stay the same when one naively changes from one point of view (yours, say) to another (e.g. the potentially alien astronauts'). New terms appear in the equations. And these new terms lead to predictions that do not match experiments.

This bothered Dutch physicist Hendrik Lorentz – a lot.

He liked Maxwell's equations (as we all should), so he wanted them to hold true, to remain the same for everyone moving at a constant speed, alien or not, fast or slow, independently of who found them first. But for this to happen, Lorentz realized that something had to give. Something that had always been taken for granted: time and space could not be universal concepts any more. They both had to depend on who measures them. And I mean the real distances you measure every day, and the real time your clock indicates.

If you already think this sounds crazy, bear this in mind: it may have been theoretical about a century ago, but it is not any more. It's been experimentally proved correct many times over. We for instance wouldn't have satellite communications had this not been figured out.

What Lorentz understood is that for Maxwell's equations to always hold true, a naive and straightforward way of shifting from one point of view to another is wrong. One needs to take relative velocities into account and transform our points of view accordingly. The mathematical transformations Lorentz found to do just that now bear his name: the Lorentz transformations. They mix space, time and relative velocities.

Which is peculiar.

Let's get you back on Earth.

You are now walking beside a road. It is a beautiful day. There's someone driving past you. It is easy to picture yourself taking the driver's place and seeing the world from the driver's seat, isn't it? You'd just need to picture yourself there, in the driver's seat, in your mind, and there you go.

Wrong, says Lorentz.

To do so properly, you'd have to adjust your time. And the distances you see.

Rather reassuringly, for the velocities we are used to in our daily lives, these adjustments are not needed. Our senses can't notice the change. They are too tiny. Even with the fastest car there is, a second on the pavement matches a second in the car. The same goes for distances: if the driver says he'll stop in 100 metres to pick you up, you'd have to walk 100 metres to get in. Velocities also still add up normally: if the driver throws a ball forwards, you'll see the ball fly at the speed the driver put behind it plus the speed of the car.

For all practical purposes, you can picture yourself driving and that would match what you'd intuitively predicted from the pavement. But none of the above holds any more when one reaches velocities that are beyond what our senses can perceive.

You are back in space. The two possible aliens are gone and it is now a spaceship you see shooting by.

It is travelling at 200,000 kilometres per second towards you.

Hoping to get a free ride, you ask its captain to stop the ship 100 metres away for you to get in (we'll assume the ship can stop dead like that). But what do you know, 100 metres for the captain now corresponds to 135 metres for you.

The distances measured from within a ship travelling at 200,000 km/s are shorter than they are when measured by you, standing outside. For future space travel this is very handy: what appear to be enormous distances for us here on Earth wouldn't be for very fast-moving space travellers.

And if you peeked at a clock inside the ship, you'd realize that its hands do not move as fast as your wristwatch's. Time, on the ship, doesn't flow as fast as outside. So not only do distances shrink at high velocities, the way time flows also changes.

The two are linked. And the reason why is, again, the speed of light.

A speed, or velocity, is a distance travelled in a given time.

And light travels a fixed distance every second: 299,792,458 metres, no matter who measures the distance.

But a metre is not a universal concept. What the alien astronauts or the ship's captain call a metre is not a metre for you, but less.

And yet, the speed of light should be constant. Always.

So, for the speed of light to always remain the same, a passing second of their time, as seen by you, has to be less than a second measured on a clock of yours. There's no way around it.

These effects are nowadays rather well known and understood. They are called *time and distance dilations* or *contractions*. Fast-moving objects are strange, from our limited human points of view. But whether one likes it or not, that's how our universe works.

And yes, velocities don't add up properly any more either.

And this gets worse and worse as one moves faster and faster, until one reaches the speed of light: c. Then we have this beautiful result: whatever you add to the speed of light, you still get the speed of light.

For instance, adding the speed of light to the speed of light gives you the speed of light.

$$c + c = c$$

Not 2c.

These are the results predicted by Lorentz's transformations, and it was a breakthrough: Lorentz had found a mathematical reason for the infamous Michelson–Morley experimental result. Velocities only add up normally when they are very small compared to the speed of light. When fast-moving objects are involved, one has to use his transformation rules. And when one does, one finds that a light beam will always move at the same speed, the speed of light, whatever the motion of its source, whatever the motion of the person who measures it. As a by-product, his transformations also said that nothing meaningful could travel faster than light. This is why any announcement of a faster-than-light signal is always met with profound scepticism by the entire scientific community.

But this calls for a word of caution: it may well be that all this is not *absolutely* correct, that distances and time intervals don't behave *exactly* as Lorentz predicted, or that something could travel faster than light. Physics is about finding the best possible model to describe reality, not about finding something perfect, or absolute. Such notions do not exist in physics (or nature as we know it). Every new finding, every new understanding, is related to the technology that can confirm, or disprove, its validity. And no technology is infinitely precise.

Nevertheless, even today, about a century later, not a single experiment has detected the slightest hint that the Lorentz transformations might be wrong.

With this said, we are now ready to understand E = mc².

CHAPTER 3

$E = mc^2$

In 1905, right after publishing his theory of fast-moving objects, Einstein linked his principle of relativity and the constancy of the speed of light to something that seemingly had nothing to do with it all: mass.

For centuries, mass was what you could measure on a scale. In our mathematical description of the world, the description Newton gave us and which is still taught in schools today, mass appears as a letter – m – which tells us how massive objects react when subjected to a force.

Mass was not supposed to change because of how an object moved. It was supposed to be fixed from beginning to end.

But pick up a stone.

Throw it so that it speeds up to, say, the velocity of light minus one kilometre per hour.

Then, as it flies, find a way to give it an extra little push that would normally get it to fly two kilometres per hour faster.

It should then find itself travelling 1 km/h faster than the speed of light.

But it can't, according to Einstein's principle (and Lorentz's transformations).

According to him, you can try to hit your stone however hard you want with a bat or a missile or anything you can think of, but you won't get it to fly faster than light.

So where does the energy from these pushes go?

Einstein's answer: into mass.

For him, the faster the stone, the more massive it becomes, the harder it is to get it to move even faster. Mass, according to Einstein, depends on speed. Even for you.

Now, when one changes the meaning of something like mass, one should be ready to face the consequences. And one of them has something to do with energy.

Energy is a rather difficult concept to define, but there are a few things we can all agree on. One is that the faster and heavier an object is that hits you, the more it hurts. So one could say that the faster and heavier an object is, the more energy it carries. This is a particular type of energy called kinetic energy, which has been known and understood since Newton. Or so we thought.

In 1905, Einstein replaced the old, fixed Newtonian mass with his new, velocity-dependent one. And he found a term within kinetic energy that did not depend on the velocity of the moving object, which is rather strange, since kinetic energy is, by definition, related to speed.

The term he found looked like mc².

Something that looked like a mass – m – times the square of the velocity of light, not the velocity of the moving object. This term appeared to be part of the energy of the stone, and because of the c^2, it was big. Really big.

But what was this m?

It is a mass that does not depend on velocity. It is the mass at rest, the mass contained in any object – be it a stone, a rock, a camel, a human, a planet or a star – as measured by someone who sees it immobile.

The meaning Einstein gave that mc² term is the following: every single object made of matter bears a fundamental energy which just comes from it existing. It is the energy every single object has for being made of, and assembled out of, the fundamental constituents of our universe.

It is the ground energy of things.

If we call the energy E, this gives:

$E = mc^2$.

It means that mass and energy are but two aspects of the same thing.

It means that you can get from one to the other.

It means that the exchange rate, if you could actually turn mass into energy, is c^2. An enormous number.

It is a direct consequence of Einstein's principle of relativity and assumed constancy of the speed of light. Einstein published this result on 25 November 1905, in a paper he entitled 'Does the Inertia of a Body Depend upon its Energy Contents?'

The question was rhetorical, of course, since he had just proved that yes, it did.

CHAPTER 4

The Fourth Dimension

When Newton was alive and well, about three centuries ago, and for as far back as I can tell, trying to picture the universe we live in involved picturing some kind of three-dimensional volume, called space, and an external clock that ticked with relentless continuity, giving us what we call time. Space and time were the same for one and all. Time was a way to tell the past from the future. Space was a way to describe shapes, and distances. To meet up with someone, for instance, you needed – and still need – to specify a place in space and a moment in time.

Were intuitive ideas like these to be correct, one could fairly easily picture shapes and clocks ticking in faraway worlds, or on fast-moving spaceships, and expect them to be as they'd be when seen from here, on Earth. Except that it is not so. Clocks do not tick the same way. Distances are distorted. Like the Lorentz transformations, say.

Lorentz transformations, however, are not very aesthetic from a mathematical point of view, making them look tricky.

Rather fortunately for us, scientists don't like tricky things. They like to simplify. Again and again. And that is what German mathematical physicist Hermann Minkowski did in the years following the publication of Einstein's 1905 papers. Minkowski, a one-time professor of Einstein's at Zurich University, found a very natural, albeit very strange, way to think about it all.

Building on mathematical discoveries made by Henri Poincaré, Minkowski realized that from the perspective of Einstein's principle, the whole lot of complicated Lorentz transformations became extraordinarily simple if one assumed that space and time were not separate entities but made up a whole. A four-dimensional *spacetime* in which the distances are calculated not by merely looking at space extensions, but time separations as well: what separates you and me right now, in Minkowski's spacetime, is given by the usual distance between us, in space, from which one subtracts the time it takes for light to travel between you and me.[4]

Distances are not merely spatial separations any more. They take into account time too. They become spacetime lengths. And to imagine how the universe changes when looked upon from different perspectives, Minkowski spacetime says you need to use . . . Lorentz transformations.

Distances and time intervals, when taken independently of each other, are not universal. Spacetime distances, however, are.

In a spacetime like Minkowski's, Maxwell's equations are automatically the same for everyone, whatever the speed (as long as it is constant).

In a spacetime like Minkowski's, Einstein's principles are fulfilled.

And it so happens that Minkowski spacetime, with its share of time and space contractions and dilations, but fixed spacetime intervals, is uncannily similar to our actual experience of reality.

Now, let me tell you something remarkable: in Minkowski spacetime, the spacetime distance travelled by something moving at the speed of light is zero. Hence light gives you a clear separation between events with a positive and negative spacetime separation.

This is what differentiates events that can have an influence on one another, and those which never will. If the distance between the two events is, say, positive, then it is OK: a message can be sent between the two. But if the distance is negative, it is impossible. No meaningful message will ever travel between the two. Two such events are, for all practical purposes, living in different causal universes.

Moreover, for events which can exchange signals, there is,

in Minkowski spacetime, no more universal notion of which event happened first: there will always be observers who would disagree on that chronology, and who would both be right, from their point of view.

Einstein's two principles and E = mc², with Minkowski spacetime and all, is part of what we today call Einstein's theory of special relativity, but the 'relativity' in this name is not there to say that everything is relative, as in 'there is no such thing as a universal truth'. Quite the opposite, actually. It is there to say that you are not supposed to use your straightforward intuition to describe space, time, distances and time intervals. What measures a metre long or lasts a second for some does not for others, and the perceived chronology of events may also differ, but the special theory of relativity allows us to relate all points of view, linking everything we know. Relativity does not mean chaos. It means that we know how to relate with others, that we know how space and time behave, even when high velocities are involved. Or aliens.

PART III:

Consequences

CHAPTER I

Antimatter

Back in ancient Greece, more than 2,000 years ago, our ancestors wondered what one would end up with were one to forever cut a piece of material in half. They called the possible end product an *atom*, which literally means 'cannot be cut'. They did not know if such a thing existed, but they gave it a name anyway, which we still use, with a slightly different definition.

Nowadays, an atom refers to the smallest constituent of a particular element. For instance, the smallest piece of gold one can have is an atom of gold. Cut it again and you still have something, but it is not gold any more. The same holds for all the elements: oxygen, iron, carbon and so forth.

We've now known for about 100 years that atoms – and matter – are aggregates of smaller things, particles that, as far as we can tell today, are made out of nothing but themselves, true atoms. But they did not inherit the name. These are called *fundamental particles.* They are quantum in nature, meaning that they obey the strange rules of the realm of the very small.

The doors of the quantum world began to open around the year 1900, when German physicist Max Planck found a formula that could explain, mathematically, why a heated black box emitted a very special radiation, called black body radiation, which nobody before him had managed to understand. In his formula, Planck introduced the idea that light was made of small packets of energy, small quanta,[5] but he did not really believe these were real. He thought about them as a mathematical trick to get the right experimental result. He did get the Nobel Prize in Physics for achieving this, though.

Five years later, during his 'extraordinary year' as it has been called by science historians,[6] Einstein showed that this was not merely a trick.

Light really is made of small packets of energy.

We call them photons today.

Einstein received the 1921 Nobel Prize in Physics for that.

In the years that followed, French physicist Louis de Broglie showed that this also applied to matter. Matter, he found, was just like light: made of small packets of energy. He, too, got the Nobel Prize in Physics for that, in 1929.

It must have been an extraordinary time to be a scientist. Can you imagine? All of a sudden, humanity had access, for the first time, both technologically and theoretically, to a new reality. The realm of the little packets of energy that make up our world, the realm of the quanta. A realm so strange and bizarre that no one understood much about it. But the maths seemed to work, and physicists Werner Heisenberg from Germany and Erwin Schrödinger from Austria discovered how these strange packets of energy, be they light or matter, could move and travel from one place to another. They figured out their equations of motion. They got the Nobel Prize in Physics for this, in 1932 and 1933, respectively.

Then something happened which made it all even better.

English mathematical physicist Paul Dirac thought that Einstein's principles should hold not just for Maxwell's equations, but in the quantum world too. In one of the greatest scientific tours de force ever, he found a way to unite Heisenberg and Schrödinger's equations with Einstein's special relativity. And he found something that is barely believable. Special relativity directly led to the idea that those small packets of energy that are called quanta could be understood as being part of some sort of sea that fills the entire universe, a sea not of aether, but of energy out of which these particles pop in, and out, of existence.

Even more astonishing: Dirac's equations said that for Einstein's principles to hold, every single charged particle that we know of should have its opposite counterpart, some sort of anti-itself, exactly the same particle but carrying an opposite charge. Dirac first figured this out for the most famous charged particles of them all: the electrons.

The electrons have a negative electric charge (photons don't have any). Were the anti-electron to exist, as Einstein's special relativity says it should, then it should have a positive charge.

Dirac wrote this in 1931.The particle was found experimentally in 1932 by American physicist Carl David Anderson. It was the first ever antiparticle to be experimentally detected – a particle of antimatter, an anti-electron.

Dirac got the Nobel Prize in Physics one year later, in 1933, and Anderson got his in 1936.

This was the first time in history that a particle that had never been seen was predicted theoretically and then found. And it was a consequence of $E = mc^2$.

Applying Einstein's principles to the quantum world led to the discovery that our universe's matter content was . . . twice as big as was thought. And I'm not talking about speculation here, as we are today with parallel worlds and extra dimensions. This is our known reality. We even use antimatter in hospitals to map our brains.

This sea of energy Dirac had found is nowadays called the quantum vacuum. It is not empty. It has some energy. And it is everywhere. According to $E = mc^2$, the quantum world can turn that energy into mass, allowing for particle–antiparticle pair creation out of the vacuum.

And if you wonder why two particles have to appear and not just one, here is the reason: it seems that nature allows for things to change in form, but not to appear and disappear just like that.

$E = mc^2$ tells us that mass and energy are but two aspects of the same thing. If there is enough energy around, it can be turned into its mass equivalent in the form of a particle.

But if a particle has a charge, for instance an electric charge, then it can't appear alone.

Before anything appears, there is no charge around. There is just energy. Afterwards, the energy is still there. Its value is exactly the same. It has just changed form. It has now become mass. There's no problem there; that's part of the beauty of $E = mc^2$. But if a single charged particle is created, there is now a charge, and there's no such thing as an equivalent to $E = mc^2$ in which a charge replaces the mass. So the spontaneous appearance of a charge cannot be understood. And that is (obviously) not acceptable. However, if another, exactly opposite charge appears as well – an antiparticle – then the total charge remains zero and we are happy. And that is what happens in nature.

Now let's summarize what we have so far. Einstein's principles led to the discovery that our universe has not three, but four dimensions (the fourth being time), and that $E = mc^2$

which, in turn, when applied to the quantum world, led Dirac to predict that antimatter exists, and that particles and anti-particles can appear out of nowhere. All these predictions have been confirmed by experiments. Energy can be turned into mass. And it is. Constantly. Everywhere. Throughout the universe.

But what about the other way round? According to $E = mc^2$, mass could also be turned into energy, so, is that correct too? The answer, as you know, is yes. But you may not know this can happen in at least two different ways, with very different consequences.

CHAPTER 2

Atomic Energy

Let's focus for a moment on these rather peculiar little fellows we call atoms. They make up all the matter we know of, throughout our universe. And they all have a similar structure: a nucleus with electrons wiggling around it.

There are 95 known atoms occurring naturally in our universe. The smallest one is hydrogen, the second smallest is helium, carbon is number 6 and oxygen is 8. Iron is 26. Gold 79.

Atomic nuclei are made of two different types of still smaller aggregates: neutrons, which do not carry any electric charge (hence their name) and protons, which carry a positive charge.

Now, an element (like carbon, or gold etc.) is characterized by how many protons its atomic nucleus contains. Hydrogen's nucleus only has 1. Helium has 2. Carbon has 6, oxygen 8, iron 26 and gold 79. You get the idea.

To assemble a nucleus, one basically only has to glue protons and neutrons together. But that is not as easy as it sounds, because protons, carrying a positive charge, repel one another. Hence one needs an impossible amount of energy to bring them close, just as one needs to be very strong to get the north poles of two magnets close to one another, let alone touch.

But here is the thing: in the world of the very small, new forces appear.

Neutrons and protons are both made up of smaller particles still, which are called quarks. As far as we today know, it ends there. Quarks are not made of smaller particles. They are fundamental. And a very special force holds these quarks together to make up protons and neutrons. It is called the strong nuclear force.

It so happens that this strong force, at very small distances, is a lot stronger than electromagnetic repulsion.

It is as if you had to fight like mad to bring two magnets' north poles close, but once you reached a critical minimum distance, they'd start attracting each other and, no matter how hard you tried, you wouldn't be able to pull them apart any more.

That's exactly what happens in atomic nuclei: there's a fight between electromagnetism and the strong nuclear force. And it is a fight to the death. There is never a tie. Who wins is all a matter of distances.

46

Now this strong force also has an effect beyond the quarks themselves, on protons and neutrons. Scientists call this a residual force. It binds protons and neutrons as it binds quarks. As long as the neutrons and protons are not too far from one another, this residual force is again stronger than electromagnetic repulsion, keeping the nuclei of pretty much all the atoms in nature safely packed. Without it, atomic nuclei would blow apart because protons would repel one another, and we wouldn't exist. But we do. So they don't blow up. And yet, since it is all a matter of distances, they do actually have the potential to blow up.

With rather dramatic consequences.

In 1939, German physicists Otto Hahn and Fritz Strassmann published a very strange experimental result. Since all atomic nuclei are made of neutrons and protons, Hahn and Strassmann had the crazy idea that if they could, say, carefully add a neutron to the nucleus of one of the heaviest atoms known, uranium, they would create a new one, heavier than uranium. Except that that is not what they got. By shooting a slow neutron at their target, they didn't get one bigger nucleus, but several smaller. This was so unexpected they waited for quite a while to publish their findings. Although still puzzled, they eventually did, and someone immediately understood what had happened. Austrian-Swedish physicist Lise Meitner worked out that by shooting their neutron at it,

Hahn and Strassmann had ruined the equilibrium within the nucleus, changed the force that was in control in there and split the atom of uranium.

She also realized that the splitting created two neutrons per neutron used, plus an enormous amount of energy, paving the way for a chain reaction.

A bomb.

It is not hard to imagine that some extra neutrons can be released when you break a big aggregate of protons and neutrons into lumps. But where did the energy come from?

Lise Meitner understood it came from $E = mc^2$.

Mass can be turned into energy indeed.

But to understand what mass became pure energy, let's have another look at atomic nuclei. What is the energy of such minute stuff made of? Two things. One is the usual mass, and one is the binding energy. According to $E = mc^2$, you can translate the binding energy into a mass equivalent. A binding mass, say. Then the real, total effective mass of all nuclei is the normal mass we are used to, plus the binding mass. That is what weighing scales actually measure. Your weight is the sum of the two.

Now, if you take a large nucleus, like one of uranium (with 92 protons and 143 neutrons), and split it, you'll get two smaller nuclei. But it so happens that the binding energy necessary to keep uranium whole is bigger than the combined binding energy of the two smaller nuclei you get. In other words, the fission of uranium leads to a loss of binding mass, which is turned into pure energy via $E = mc^2$. This is the source of energy that is radiated by radioactive materials, harvested in atomic reactors or released in atomic bombs. This is why $E = mc^2$ is so famous. But this only applies to big nuclei.

For small ones, it is actually the other way round.

If you take two small nuclei and fuse them to create a bigger nucleus, the binding energy of the big nucleus you get is smaller than that of the two nuclei you started with. This means that for small atoms, you lose mass not by splitting them, but by fusing them. The exact opposite of the fission.

This is what happens in star cores.

Stars fuse small atomic nuclei in their cores to create larger ones, and the mass that is lost in the process, once turned into energy via $E = mc^2$, is what makes them shine.

$E = mc^2$ explains how stars generate energy by fusing atomic cores to create the matter we today are made of. As they do so, they create, out of hydrogen and helium, larger atoms, up to iron. Beyond iron, the fusion requires energy and does not release any, so that all the elements that are around us which are heavier than iron, like gold, are not forged during a star's life, but at its death, when part of the immense energy released by its explosion is used to create those heavy atoms we see around us today.

E = mc² is the reason why matter is created, and destroyed, in our universe.

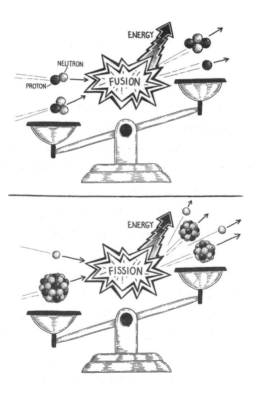

Conclusion

One of the greatest physicists of the twentieth century, Richard Feynman, whose understanding of $E = mc^2$ inspired generations of scientists (including me and this book),[7] once compared the universe to a great chess game being played. Theoretical physics, then, is all about finding the rules of the game. And we have found some. Not all of them, mind you, but some nonetheless. We can be very proud of ourselves for that. We are, as far as we know, the first species ever to achieve this feat (at least on Earth).

And it so happens that to express these laws in a way that allows for predictions to be made about unknowns, we have also found a language: mathematics. Why it is so, I cannot say. I doubt anybody can. It may even be that one day we will find another, even better one, but, as of today, mathematics is the only one we have that actually works.

Within this language, a particular set of 'words' or

'sentences' have special importance. We call them formulas, or formulae. They are, to approximately quote Feynman again, the memory of which laws of nature we have found so far.

And one of these formulas stands out:

$E = mc^2$.

It is the best known of them all.

It says that mankind has unravelled an extraordinary secret of nature: a deep link between matter and energy.

In the future memories of our species, it encompasses the idea that space and time are one: spacetime; that light always travels at the same speed in a vacuum; that nothing can travel faster than that while carrying a message of any sort; that the laws of nature are the same for everyone travelling at constant speed; and that distances and time intervals are not universal but depend on who measures them.

That all these ideas, and more, are contained in such a tiny formula is astonishing.

It says that we know how nature creates energy out of matter and matter out of energy, and that we could use it for our own benefit. We have just started doing so, as a species, and we have a lot to learn, and a lot of growing up to do, to use it wisely and safely, unlike what has been done in the past. This knowledge may be the key to sustaining our civilizations in the long term and maybe even one day reaching faraway planets orbiting faraway stars, with spaceships travelling fast enough to shrink distances and travel times so much that one could get to new worlds within a lifetime.

But even more extraordinary than all this, maybe, is the man behind it. Albert Einstein.

He embodies the triumph of pure thought. Through mere thinking, believing his two principles superseded any previous common wisdom, he opened up new realms of nature for us.

And ten years after publishing his theory of special relativity, he extended his principle of relativity to objects whose velocities are not constant any more. It became a theory of gravity. He called it not the special, but the general theory of relativity. All we know about our universe as a whole stems from it.

The faith he had in his own reasoning powers is astonishing. Many people believe they are right when stating an idea that drifts away from accepted wisdom. The only difference with Einstein is that he actually was.

Bibliography

Albert Einstein, 'Zur Elektrodynamik bewegter Körper', *Annalen der Physik*, vol. 17, no. 10, 30 June 1905, pp. 891–921 (where E = mc² was posited)
  Albert Einstein, 'Ist die Trägheit eines Körpers von seinem Energieinhalt abhängig?', *Annalen der Physik*, vol. 18, no. 13, 1905, pp. 639–41 (where he shows that light is made of packets of energy)
  Albert Einstein, 'Über einen die Erzeugung und Verwandlung des Lichtes betreffenden heuristischen Gesichtspunkt', *Annalen der Physik*, vol. 17, no. 6, 1905, pp. 132–48

Here is an English translation of all Einstein's 1905 papers:
  Albert Einstein, John Stachel, Roger Penrose, *Einstein's Miraculous Year: Five Papers That Changed the Face of Physic*, Princeton University Press, 2005

And if you want to know more:

Richard Feynman, Robert Leighton, Matthew Sands, *The Feynman Lectures on Physics: The Definitive and Extended Edition* (2nd ed.), Addison-Wesley, 2005 [1970] (a transcript of one of the best series of physics lectures ever given)

Richard Feynman, *Six Not-So-Easy Pieces, Einstein's Relativity, Symmetry and Space-Time*, Addison-Wesley, 1997

But beware, the above two Feynman books are rather involved for beginners and use equations.

For a more general introduction to everything we know about our universe, if I may, I would suggest this:

Christophe Galfard, *The Universe in Your Hand*, Macmillan, 2015

and Bill Bryson's masterpiece of popular science:

Bill Bryson, *A Short History of Nearly Everything*, Doubleday, 2003

Notes

1 The Earth has only one moon, called the Moon, but Jupiter has at least 54. Four of them are large and rocky. Io, one of the four large and rocky ones, is the one closest to Jupiter.

2 Whether it is through air or a solid or a liquid, you can hear a sound, but not in outer space. There's no sound where there's no matter for it to travel through.

3 But light can (and does) slow down when not in a vacuum. French physicist Serge Haroche received the 2012 Nobel Prize in Physics for managing to stop light.

4 There are actually squares and square roots involved, but the idea is the same.

5 Quantum (plural quanta) means 'little packet' in Latin.

6 They actually call it the *annus mirabilis*, which is Latin for extraordinary year.

7 See Feynman's *Six Not So Easy Pieces* in the bibliography.